Johannes Gutenberg

Man of the Millennium

Aaron J. Keirns

Howard, Ohio

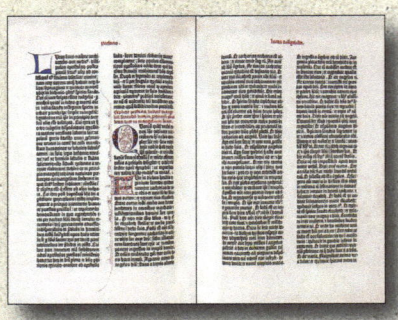

In commemoration
of the 550th anniversary of the
death of Johannes Gutenberg.
1468 ~ 2018

To my family.

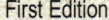

First Edition
Johannes Gutenberg: Man of the Millennium
ISBN 978-0-692-10418-7
© 2018 Little River Publishing
Book Design by the Author. All rights reserved.

No part of this book may be reproduced in any manner without written consent of the publisher, with the exception of brief excerpts for review or teaching purposes.

The information printed herein was obtained from a variety of sources, including old books and documents of unknown provenance. We have attempted to verify all information but, due to the historical nature of the subject matter, accuracy is not always certain. Due to the age of some images, we were not always able to determine the name of a creator or copyright holder.

ONTENTS

Preface v

1 *Man of the Millennium* 1

2 *A Printing Revolution* 13

3 *The Power of Books* 35

> *Postscript: Gutenberg vs. Coster* 41
> *About the Author* 42
> *Acknowledgments* 42
> *Glossary* 43
> *Colophon* 44
> *Index* 45

Detail of the Gutenberg statue in the public plaza of Mainz, Germany.

Photo by the Author

Preface

❖

This book is an introduction to the life and work of Johannes Gutenberg, the man credited with inventing the printing press around 1450.

Gutenberg has been called the "Man of the Millennium" by Time-Life Magazine and others. He was an inventor and innovator who developed the first practical mechanical printing system for books. His achievement helped launch humanity into a new age of information, education and enlightenment for the masses.

There are many individuals who deserve the honor of being named Man (or Woman) of the Millennium. Over the past 1,000 years every field of endeavor has produced exceptional men and women whose contributions changed the course of history. However, Gutenberg is somewhat unique. His work enabled the mass distribution of the printed word for the first time. Books changed everything. Like seeds scattered across the world, they sprouted new ideas and discoveries that have affected virtually every aspect of modern life. Even in our electronic age, the printed book is still a powerful force.

This book is by no means a comprehensive account of Gutenberg's life or work. It is merely an introduction to the man and his machine, as well as a brief look at the Printing Revolution and the power of books. I spent several years working in the printing industry at a time when printing technology was changing dramatically. Consequently, I find Gutenberg's story particularly fascinating. I enjoyed creating this little book and I hope you enjoy it too.

AJK

The Many Faces of Johannes Gutenberg

There were no portraits of Gutenberg made during his lifetime. These portraits, and many others, were drawn by artists who never saw him. And though we don't know what he really looked like, the artists all seem to agree that he had a forked beard and fuzzy round hat. Consequently, his distinctive beard and hat have become iconic symbols by which we identify the man.

CHAPTER 1

Man of the Millennium

"Whatever the world is today, good and bad together,
that is what Gutenberg's invention has made it:
for from that source it has all come."

~ Mark Twain ~

utenberg. It's a name you may vaguely remember from your high school history class as the man who invented the printing press.

His full name was Johannes Gensfleisch zur Laden zum Gutenberg. No one knows what he looked like but he is usually depicted with a forked beard and fur hat. It's an image that has been replicated in various versions for more than 500 years, turning this 15th century craftsman into an almost mythical figure.

Gutenberg was indeed a real man who was born in Mainz, Germany, around the year 1400 and died in 1468. His relatively simple inventions and innovations have impacted the world in ways he could never have imagined.

To better understand why this was such an important and world-changing accomplishment, let's take a look at how books were printed before Gutenberg's invention.

One Book at a Time

Prior to Gutenberg's printing press every book had to be lettered by hand, one page at a time. This was a slow and expensive process. It could take a monk or scribe several months, even a year or more, to create one copy of a book.

Saint Jerome in the Scriptorium

Until Gutenberg developed his mechanical printing process, books had to be copied by hand. Since most of the books at the time were bibles, prayer books or books on religious topics, they were usually copied by monks or scribes. The room where they were copied was known as a scriptorium.

Image, Wikimedia Commons

Bibles were so rare and valuable that churches encased them in leather and iron bindings and chained them down. Most of the books that existed at the time were religious texts. The general populace had no books nor could they have read them anyway since most commoners were illiterate. Gutenberg's printing technology generated a flood of 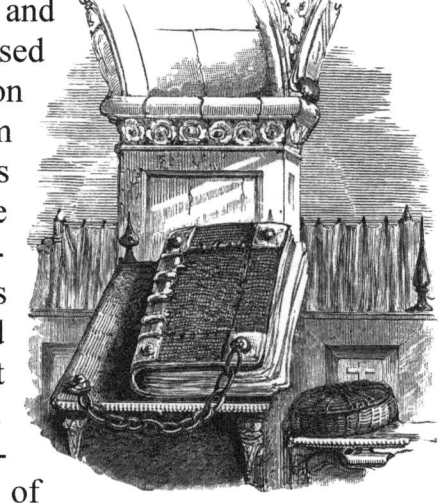 new books on a wide variety of topics, spreading information, ideas, literature and literacy across Europe and around the world.

Movable Type

Although he is commonly credited with inventing the printing press, (or even printing itself) it would be more accurate to say that Gutenberg invented the first practical printing system that utilized movable type.

Movable type refers to individual letters cast in metal that can be assembled to create words and entire pages of text. The assembled type is clamped into a frame, coated with a thin layer of ink and pressed onto paper to make an impression. Gutenberg's movable type was the key that unlocked the potential of the printing press.

Gutenberg didn't invent the press. Wooden screw presses were already in use for making paper and for pressing grapes to make wine. As it happened, Gutenberg was born in wine country. Mainz, his hometown, is considered the wine capital of Germany. Even today, the hillsides of southwestern Germany are blanketed with miles and miles of vineyards. Gutenberg was no doubt familiar with the large wooden screw presses used in wine production. He simply adapted the press design to meet his needs.

The concept of movable type was not new in Gutenberg's time. The Chinese and Koreans had been experimenting with movable type for centuries using type made of wood, ceramic and cast bronze.

For a variety of reasons, none of these previous attempts proved practical for printing books. Carving small letters in wood was extremely time-consuming and the resulting type was not very durable. Ceramic type was fragile and not suitable for mass production. The Koreans experimented with cast bronze type more than half a century before Gutenberg but with little success. The Chinese faced a unique problem. Their language consists of thousands of characters, making it very difficult to carve or cast a sufficient supply of "type" for printing something like a book.

Although Gutenberg didn't actually invent movable type, he did invent the first practical system for manufacturing mass quantities of precision, durable metal type.

Gutenberg's genius was in his ability to both invent and innovate. He invented techniques and processes based on the innovative use of existing technology and materials. Today we might characterize his work as a synthesis. He combined simpler technologies and processes to create a new and more complex system. The resulting machine was truly greater than the sum of its parts.

Renaissance Man

Gutenberg was a Renaissance man in every sense of the word. He lived during the period in Europe known as the Renaissance (14th–17th centuries) and he had skills, talents

A bronze statue of Gutenberg towers over visitors in the public plaza of Mainz, Germany.

Photo by the Author

A 19th century depiction of Gutenberg in his shop.

and knowledge in a variety of areas.

He had experience as a goldsmith, which helped him envision a practical way to cast metal type. He experimented with metallurgy in order to create an alloy for making type that was durable, precise, and didn't shrink on cooling.

He also developed a new ink. The water-based inks available at the time were made for hand-lettering documents and were too thin for mechanical printing. After experimenting with different ingredients he came up with a thicker oil-based ink.

Gutenberg combined these innovations, and others, to create an integrated system for duplicating books mechanically. It was a giant leap in technology that began a rapid spread of information and ideas that continues to this day.

Craftsman & Artist

Gutenberg was an excellent craftsman. The precision of the type he manufactured and the quality of his printing show that he had an eye for detail and exceptional skills. His work also shows something else, something not often emphasized. He was an artist. The type he made and the books he printed are true works of art.

His type is beautifully proportioned. The shape of each letter is carefully sculpted to work in harmony with every other letter, showing an intuitive sense for the nuances of letter design.

The hand of the artist is also evident in the pages he printed with his type. In the next chapter we will look at the famous "Gutenberg Bibles" he printed. The pages of his bibles have an elegant simplicity. The width and placement of the columns of text show an artistic sensibility for balance and symmetry. Each column is remarkably consistent within itself and in relation to the other columns. It is powerful design. Even for those of us who cannot read Latin, the pages of Gutenberg's bibles communicate the importance of their content.

Although we tend to think of Gutenberg primarily as an inventor, innovator and craftsman, the artistic quality of his work is equally significant. His high standards set the bar for every typographer and printer who followed.

Man of Mystery

Much of Gutenberg's life remains a mystery. Personal records from the 15th century are scarce. A lot of what has been written about his life is speculation based on bits and pieces of information gleaned from court documents and other records. He printed other things besides bibles but since he didn't put his name on anything he printed, there is very little that can be positively attributed to him.

During the printing of his bibles, Gutenberg had a dispute with his partner/financier, Johann Fust, about the repayment of a loan from Fust. The case went to court and Fust won, effectively bankrupting Gutenberg. Fust then took possession of Gutenberg's printing equipment and finished printing the bibles.

Although Gutenberg didn't get to finish his bibles, he continued working in the printing business for several years, printing or supplying movable type to other printers.

In 1465 Gutenberg's achievements were recognized by the archbishop of Mainz. Gutenberg was granted an annual pension of grain, wine and clothing, which allowed him to live in relative comfort until his death in 1468.

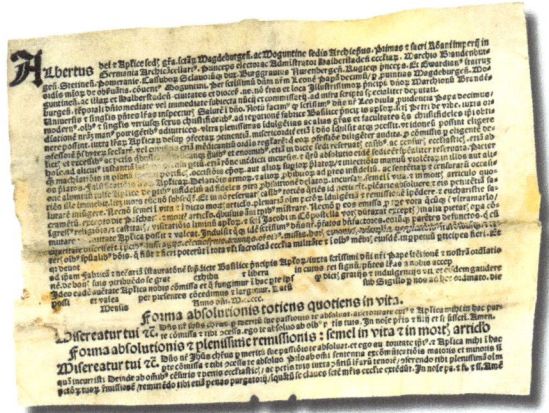

Letter of Indulgence

In Gutenberg's time, the Roman Catholic Church sold "indulgences," which granted the purchaser remission from his sins or a reduced time in Purgatory. This letter of indulgence, dated 1455, is attributed to Gutenberg's press. Indulgences were a lucrative source of income for the Church. This practice outraged Martin Luther who included it in his "95 Theses," a document that became a catalyst for the Protestant Reformation.

Image courtesy of the Gutenberg-Museum, Mainz

Chapter 1: *Man of the Millennium*

A few of the many postage stamps honoring Gutenberg.

Gutenberg's Hometown

The city of Mainz is located on the Rhine River in southwestern Germany. It was a Roman military post in the 1st century B.C. and is now the capital of the state of Rhineland-Palatinate. Many of its historic buildings were heavily damaged during WWII but have since been restored. The public plaza and market in the city's historic center is a popular tourist destination.

Left
A bust of Gutenberg at the entrance to the Gutenberg Museum is one of the few depictions showing him without his characteristic bifurcated beard.

Below
The beautiful Market Fountain in the Mainz public plaza, built in 1526, is the oldest Renaissance fountain in Germany.

Chapter 1: *Man of the Millennium*

Gutenberg Museum in Mainz

Mainz Public Plaza

Photos by the Author

Early Printing Press

A print shop as it would have looked in 1520. The pressman pulls the lever, pressing the paper against the inked type. The man at the left is holding leather-covered inking pads, preparing to re-ink the type for the next impression. The man on the right is setting a line of type in his composing stick.

CHAPTER 2

A Printing Revolution

"Books are the building blocks of civilization."

~ STEPHEN FRY ~

ike all inventors and innovators, Gutenberg wasn't working in a vacuum. He drew on the knowledge and experimentation of his predecessors and contemporaries.

He made use of existing technology, equipment and resources. He likely consulted with others while trying to work through technical problems. Even though inventors sometimes have "Aha!" moments, success usually comes only after a long process of trial and error. Gutenberg succeeded because he persevered long enough to overcome the technical obstacles that stood in his way.

Mechanical Printing

The idea of reproducing books mechanically wasn't unique to Gutenberg. At about the same time as he was printing his bibles, other artisans were producing books known as "block books." They were called this because each page was printed using a single image carved into a block of wood.

A page of a block book often contained an illustration and some text, all carved as one. The method wasn't practical for text-intensive books like bibles but it was sufficient for small prayer books and the like. They were essentially picture books with some integrated text. The pages were typically printed on one side only. These relatively inexpensive alternatives to typeset books were designed to meet the needs of a population that was mostly illiterate.

Gutenberg had a loftier goal. He wanted to mechanically reproduce bibles that were as good or better than the most beautiful hand-lettered bibles in existence. The technical challenges he faced were substantial. It would take years of experimentation for him to achieve success. The system he eventually invented for casting and printing with movable type was ingenious.

Type Casting

Gutenberg knew intuitively that his movable type had to be made of cast metal. It was the only material that made sense for reproducing the thousands of individual pieces of type required to print a book.

He needed a way to cast the individual letters quickly, inexpensively and with great precision.

The casting metal had to melt at a relatively low temperature and not shrink during the cooling process. The finished type had to be durable, reusable and hold its shape under the pressure exerted by the press. These were all difficult challenges.

He experimented with different metals until he came up with an alloy of lead, tin and antimony that met his requirements. He also had to figure out a way to make molds and develop a process for casting the letters efficiently.

In order to cast an object in metal, a negative impression of the object has to be made in some kind of suitable material, into which molten metal can be poured. This is known as a mold (also spelled mould) or matrix. Clay and sand had been used for making molds for centuries but it wasn't suitable for casting thousands of tiny pieces of intricate type.

Gutenberg needed a tough and dense mold material that could repeatedly reproduce very small letters with sharply defined edges.

His solution was to create a matrix block made of copper. Copper is a relatively soft metal but dense enough to create a mold that would faithfully reproduce even the most delicately designed type. All he needed to do was somehow make a negative impression of each letter in the copper.

He was familiar with hard metal punches that were used for stamping coins and other objects. By shaping the end of such a punch into an individual letter, he could then drive it into a small block of copper to create a negative impression. The resulting matrix block provided a mold for casting as many duplicates of the letter as he needed.

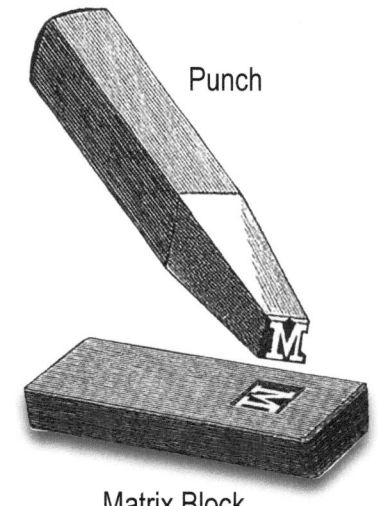

Punch

Matrix Block

He had to sculpt a punch for each letter of the alphabet, upper and lower case, as well as numbers and all necessary punctuation marks. He also made several ligatures, which are two or more letters attached together as one ("æ" for example). This was a device that scribes used in hand-lettering for aesthetic (æsthetic) reasons, and to save space and speed up their tedious work.

Hand-Held Type Caster

Now that Gutenberg had the right metal and good molds, he had to devise a practical way of precisely pouring the metal into the molds to create thousands of pieces of type.

His solution was to design a hand-held casting device. His hand-caster was a precision tool made of wood and steel. It was perhaps his most ingenious invention.

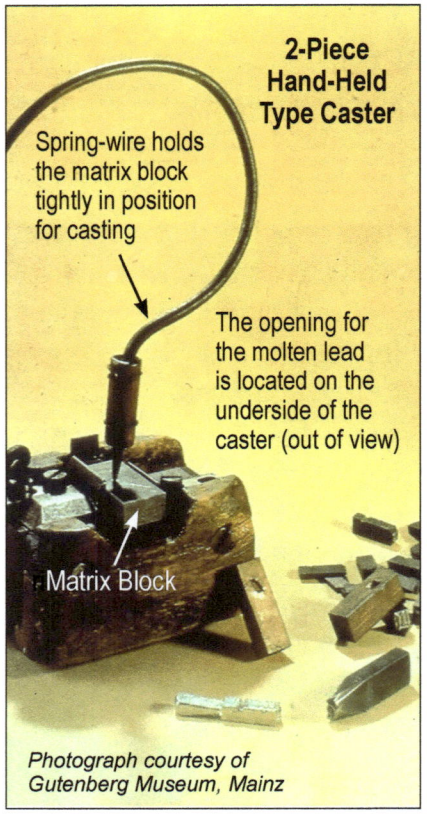

2-Piece Hand-Held Type Caster

Spring-wire holds the matrix block tightly in position for casting

The opening for the molten lead is located on the underside of the caster (out of view)

Matrix Block

Photograph courtesy of Gutenberg Museum, Mainz

With a matrix block inserted, the caster would create a slender bar with the letter protruding from its end *(see illustration on page 17)*.

This bar or "body" of the type gave all letters a consistent height, which was crucial for producing an evenly-printed page. When the type was assembled into columns, the surfaces of the letters needed to be flush (on the same plane). Low or high spots would cause the ink to transfer to the paper unevenly making some letters print too light or too dark.

Gutenberg made his hand caster in two halves that could be opened and closed quickly with a spring-wire mechanism. He held the closed caster in one hand and poured-in the molten lead with the other while simultaneously thrusting the caster upward to force the lead down into the matrix. The lead cooled almost instantly. He then released the spring mechanism, separated the two halves of the caster, removed the type, closed the caster and started all over again. The process was fast and efficient. Using the caster, one man could produce thousands of pieces of type per day.

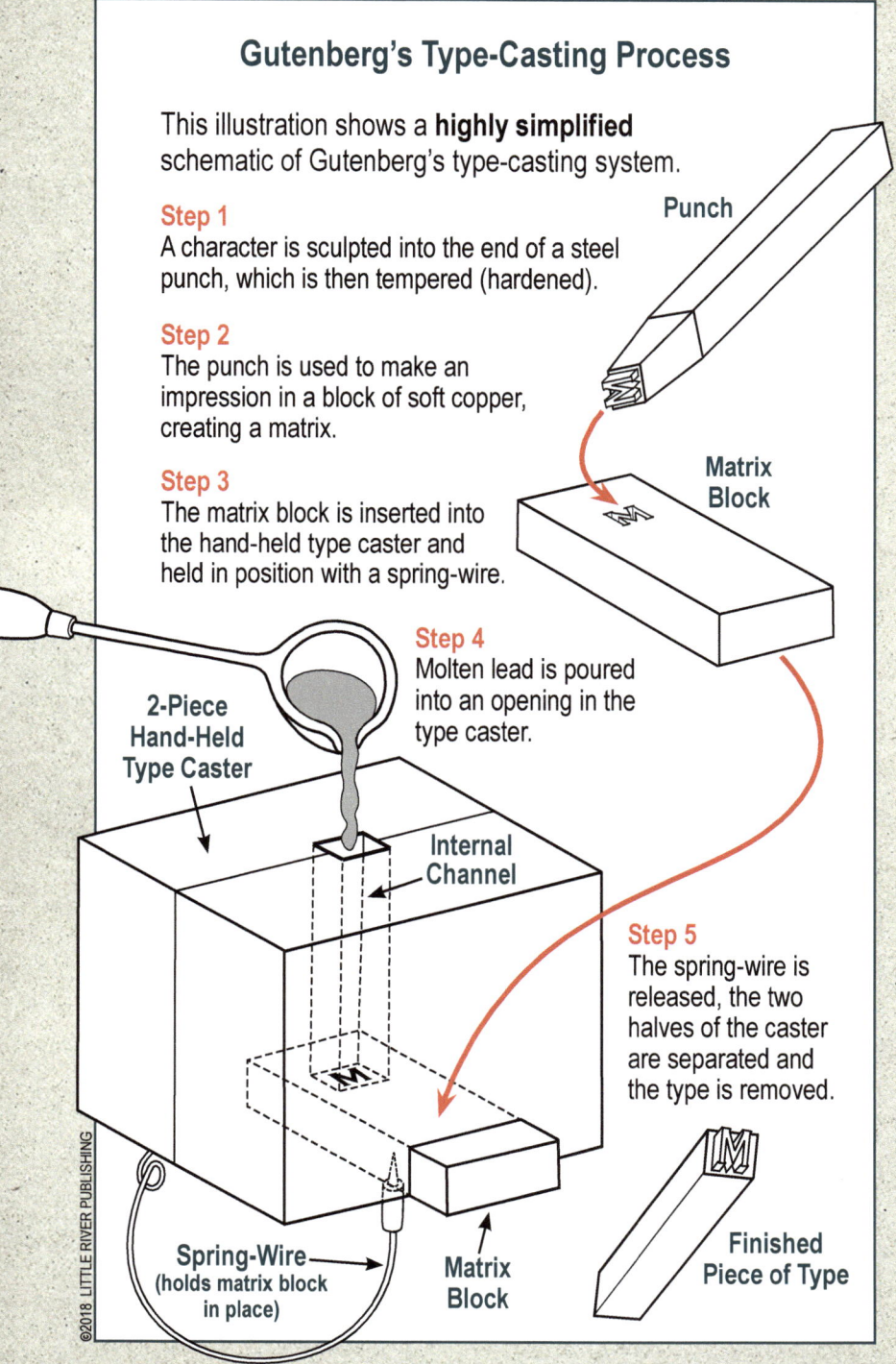

Compositing or Typesetting

After the type was made, it was placed in shallow divided cases with separate bins holding quantities of each character. The type was assembled for printing by a compositor or typesetter. Standing in front of the type case, the compositor picked out the individual letters needed to build the words, sentences and paragraphs he wanted to print. There were generally two cases of type within his reach: a lower case which held the small letters and an upper case holding the capital letters. This is where the terms "lowercase letters" and "uppercase letters" originated.

The compositor held a "composing stick" in his left hand and picked out the letters with his right hand. He placed each letter in the composing stick in the correct sequence for the words he was spelling. He assembled the letters from left to right and upside-down. Essentially, the compositor created a mirror image of the text he wanted to print. If he were to assemble the text in right-reading order, it would print backwards on the paper.

Composing Stick
Wikimedia Commons

After a few lines of type were assembled in the composing stick, the compositor transferred the block of type to a frame. He continued this process until the entire page of text was assembled in the frame exactly as he wanted it to print on the page. He then clamped the frame of type into the press, ready for inking.

Replica of an early printing press on display at the Gutenberg Museum. This kind of printing is often referred to as "letterpress."

Image courtesy of the Gutenberg Museum, Mainz

Gutenberg's Press

There is no way to know exactly what Gutenberg's printing press looked like. There are several woodcut illustrations from the 15th century that show printing presses and it's likely that his press was similar to those. He probably had to modify his press several times as he fine-tuned his printing process.

The presses from that time period had heavy wooden frames. They had to be able to withstand the pressure placed on them by the lever and screw mechanism. It took a lot of force to press the paper firmly against the type. Printing presses remained similar in design for hundreds of years after Gutenberg. Even as late as the 19th century, when presses were made of cast iron, the overall design was reminiscent of the earlier wooden presses.

Ink & Paper

As mentioned previously, Gutenberg formulated a new oil-based ink for his printing system. The ink contained resin and soot, and was thick enough to stick to the metal type yet thin enough to leave a sharp impression on paper or vellum. This was a critical part of his printing success. He no doubt had to try a variety of formulations before coming up with an ink that met all of his requirements. It's another example of Gutenberg's ability to adapt and innovate.

In Gutenberg's time, paper was handmade, relatively expensive and not readily available in large quantities. His printing press changed all that. As printing became more common, paper mills sprang up to fill the need. Vellum was also a common material used for printing at the time. It was made from the skins of calves. Because of the labor and time involved in preparing the calfskin, vellum was much more expensive than paper.

Gutenberg's printing system was multifaceted. All of the materials involved, metal, ink and paper, had to work in unison with the mechanics of the press in order for his printing system to work properly.

The Gutenberg Bible

As if his invention of movable type wasn't enough, Gutenberg is equally famous for the exquisite books he printed using his movable type. The Gutenberg Bibles, as they have become known, are perhaps the most valuable books in the world.

Gutenberg Bible

There was a growing demand for bibles in Gutenberg's time. The Roman Catholic Church was wealthy and powerful. It wanted to spread its influence across Europe and needed bibles for its many churches. A hand-lettered bible could take a year or more to produce, and scribes introduced unintentional errors. There were also inconsistencies in the content of bibles produced by different churches. The Church needed bibles that delivered a consistent doctrine everywhere. Gutenberg's mechanical reproduction system offered a solution to this problem. Every page of text was exactly the same. And because they were produced in quantity, his bibles would be much less expensive and more readily available.

The bibles Gutenberg printed were large. The page size is about 12" x 16." Each bible has roughly 1,282 pages. It is estimated that he printed about 180 copies of the bible, which means he had to print nearly 231,000 pages.

IMPRES
Poteſt vt vna vox capi aure plurima:

Print Shop

In Gutenberg's time, the ability to reproduce exact duplicates of books was an astonishing accomplishment. Some people thought it might be the work of the Devil.

Devil or not, Gutenberg's printing technology quickly spread, helping to increase literacy across Europe.

This engraving, *ca.* 1600, is titled *New Inventions of Modern Times*, and depicts a scene in a "book mill."

He printed 150 of the bibles on paper and another 30 copies on vellum. Vellum is beautiful and durable but very expensive. It is estimated that it took the skins of roughly 140 calves to make one Gutenberg Bible. If this is accurate, it took around 4,000 calves to make the 30 copies.

Scholars believe that Gutenberg may have had as many as six presses and 20 craftsmen working on the bibles simultaneously. It may have taken as long as three years to finish the 180 bibles.

Only about fifty of the bibles survive today and many are incomplete. Most are in museums. If a complete bible were to come on the market today, it is estimated that it would sell for $25-35 million. Individual pages from a Gutenberg Bible can sell for $100,000 or more.

When Gutenberg decided to print the bible, he based his type design and page layout on existing bibles that had been lettered by hand. The precision he was able to achieve is remarkable. Even though his bible was the first mass produced book, it was a masterpiece.

He was able to precisely adjust his letter and word spacing so that the columns of text align on both margins. This is known as "justified type," and is not an easy thing to do. Even today, when our books (including this book) are laid-out on sophisticated computers, the spacing of the justified type can sometimes look awkward.

Today, computers control the width of a column of digital type by adjusting the spaces between letters and words, and by the use of hyphenation. Depending on the particular type style, type size and column width, some lines of type can look a little too spaced out and some a little too tight. Computers hyphenate words in order to help even-out the text. Too many hyphens, however, can be a distraction for the reader.

Illumination & Binding

When Gutenberg finished printing a bible, it was still a long way from being a completed book. Every bible needed to pass through the hands of a couple of other skilled artisans before it was truly finished.

Gutenberg printed only the black text. He left blank areas in some columns of text to accommodate illuminated (ornate) capital letters and other colored text that would be inked-in by hand later. Illumination was common in hand-lettered bibles and Gutenberg wanted his books to look like the bibles people were accustomed to.

After all the black text had been printed, the loose pages were given to an artist who added ornamental capitals, flourishes and colored lettering throughout the book. Every artist or scribe had his own style. Consequently, each Gutenberg bible is unique. Some artists drew only simple capital letters while others added elaborate flourishes, gold leaf and vignettes with miniature scenes. The same page from two different bibles can look dramatically different.

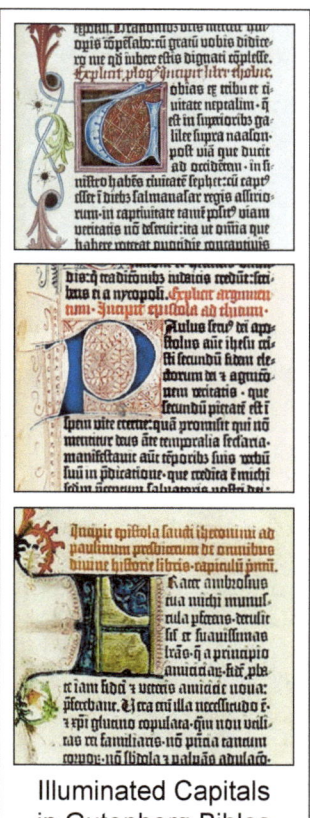

Illuminated Capitals in Gutenberg Bibles

After the illumination was added, the bible was taken to a bindery. The binder trimmed, stitched and glued the pages into a leather binding made to the customer's specifications. A book lover might have all of his books bound alike in order to keep his personal library coordinated.

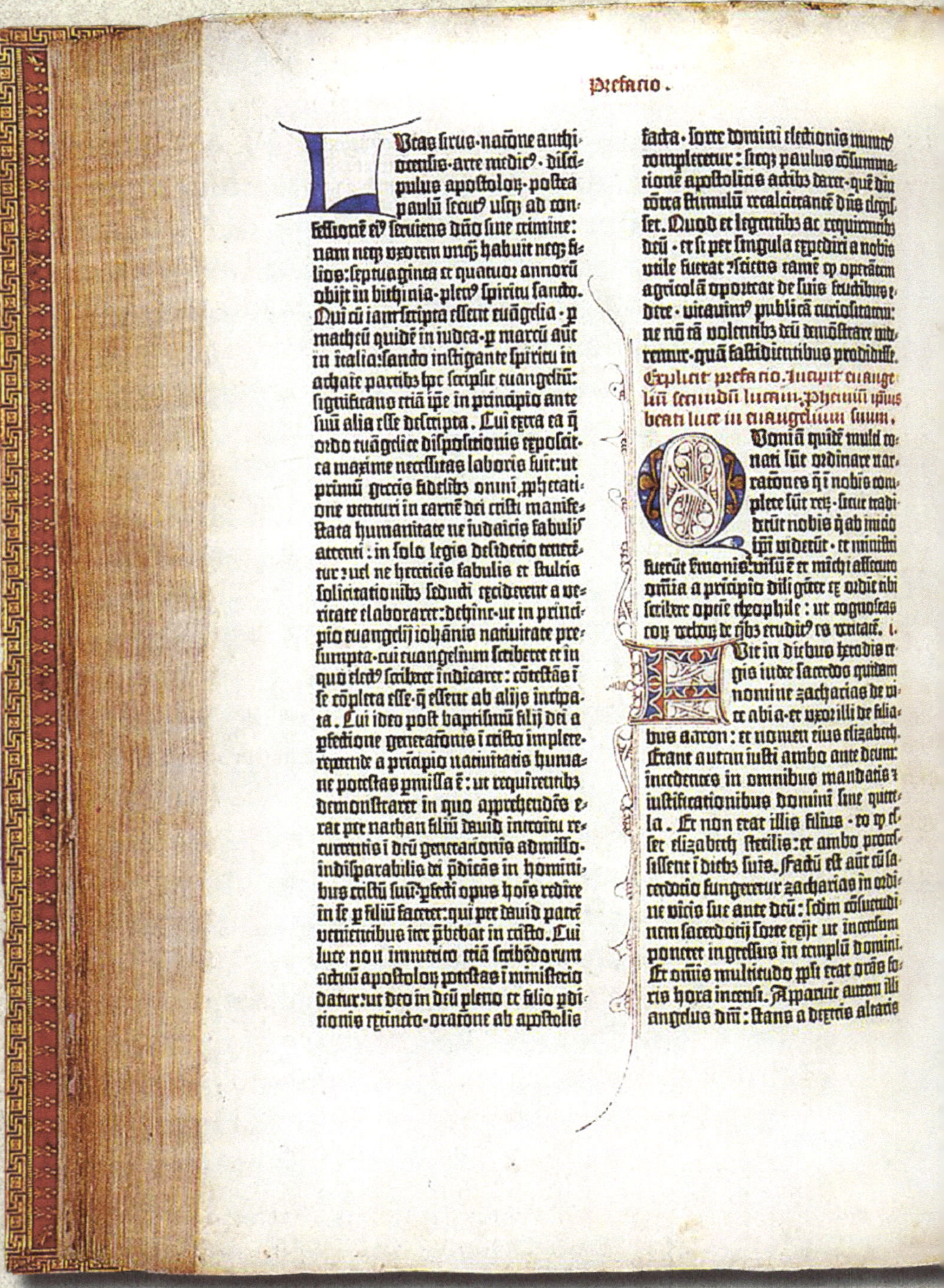
Prefacio.
Explicit prefacio.

Chapter 2: *A Printing Revolution*

Inset showing a closeup of Gutenberg's type design.

Gutenberg Bible

A two-page spread from the Gutenberg Bible (Shuckburgh Copy) on display at the Gutenberg Museum. The pages are from the beginning of *The Gospel According to Luke*.

Gutenberg printed only the black columns of text, leaving blank areas for the illuminated letters, which were inked-in by hand later, along with the rubrication (red lettering).

Bible image courtesy of the Gutenberg-Museum, Mainz

Printing Evolution

Gutenberg's movable type and printing press were so well designed that printing didn't change much over the next 400 years. It wasn't until the 19th century that printing press design started to evolve in earnest. Machines began taking over processes that had always been done by hand. By the late 19th and early 20th centuries, machinery was developed that could cast and set type. Steam and electricity began to replace muscle power and the printing process became faster and more automated.

One of the most significant changes to printing came in the 1880s with the invention of the Linotype machine. The Linotype enabled one man to sit at a keyboard and set as much type as five men could set by hand. It was a quantum leap in speed and efficiency, and made it possible for weekly newspapers to become daily newspapers. Thomas Edison called it the "eighth wonder of the world." The Linotype was arguably the most important advance in printing since Gutenberg. Remarkably, the inventor of the Linotype, Ottmar Merganthaler, was born about 100 miles south of Gutenberg's birthplace.

The linotype operator had a 90-character keyboard similar to a typewriter. As he typed, the machine assembled matrices (plural of matrix) in a line to form sentences. The machine pumped molten lead into the assembled matrices, casting the entire line as a single piece or "slug."

The next major change came in the 1960s with phototypesetting. Type was produced photographically and used in conjunction with light-sensitive printing plates. This essentially eliminated the need for lead type.

Then in the 1980s computers started changing everything in the world of printing. Desktop publishing, digital type, laser and inkjet printers and high speed computerized presses are just some of the more notable innovations.

By the early 2000s, a technology called print-on-demand (POD) publishing became viable. With POD, a compact machine can print, collate and bind an entire softcover book in only a few minutes. This major printing innovation has changed the nature of book publishing. Books can now be printed in small numbers (even a single book) efficiently and economically. This new technology has empowered authors to become self-publishers. Gutenberg put books in the hands of the masses for the first time. Now POD is putting publishing in the hands of the masses. It's an exciting time for authors, publishers and readers.

The End of the Printed Book?

As we have seen, book production technology has advanced dramatically over the past 500 years. Gutenberg wouldn't recognize a modern printing press. But of all the advances that have been made, there is one that would surely mystify him: e-books.

Electronic books (called e-books or digital books) exist as pixels on an electronic display. They are read using a dedicated reading device like Amazon's Kindle or Barnes & Noble's Nook, or on a personal computer, laptop, tablet or smartphone.

E-books have some great advantages over printed books. They take up almost no space. An e-book reader can liter-

Wheel of Books

This 16th century machine known as a "book wheel," was like a Ferris Wheel for books. It facilitated the reading of several heavy volumes in one location. Imagine what this fellow would think if he knew that today we have digital devices that enable us to hold thousands of books in the palm of one hand.

ally hold thousands of books. E-books are searchable and can contain images, video and sound. Yet even with these amazing features they have not replaced printed books.

Do e-books signal the end of the printed book? No. At least not any time soon. History shows that we humans are slow to change our ways.

In the 19th century when photography was introduced, people were certain it would lead to the end of drawing and painting. After all, why would we need artists if anybody could just point a camera at something and capture a perfect image? As we know, photography found its own niche to fill and has become a respected art form all its own. Meanwhile, drawing and painting are still alive and well. The advent of photography actually helped free artists from their duty to imitate visual reality. Impressionism was to some degree a reaction to the static reality of the images being produced by the emerging technology of photography.

In the early twentieth century, some predicted that the advancing technology of television would mean the end of radio. Why would anyone want only sound when they could have sound and picture? But as we can see (and hear) radio is more popular today than ever. The two technologies coexist, each offering particular advantages.

As technologies change, we adapt. E-books are a wonderful new technology that can do things printed books can't. But like photography and television, this new technology is filling a niche that didn't previously exist. Printed books and e-books each have distinct qualities and values in the marketplace.

That being said, there are some specialized kinds of books that will likely be replaced by e-books. Printed encyclopedias, dictionaries, directories and phone books are becoming 21st century dinosaurs.

Printed books transcend time and space. They have survived and prospered for centuries in diverse environments and cultures. Like the wheel or the lever, books have a simplicity of form and function. We like how they feel in our hands. They have weight, volume and texture. Like magic, we can open their doors and step into other worlds. E-books, on the other hand, are pictures of books; two-dimensional imitations of the real thing.

Books have been an integral part of our lives for way too long to suddenly become obsolete. They form the foundation of our collective knowledge. Writer Stephen Fry quipped that: "Books are no more threatened by Kindle than stairs by elevators."

Information Unplugged

Most of us are comfortable with the straightforward, fail-safe technology of printed books. They are intuitive to operate and don't require a power source, software or tech support. Books are information unplugged.

With a printed book there is no need for a user interface. The user is the interface. We can dog-ear the pages and write in the margins. One page doesn't go away when you turn to another page. It's all there, all the time. Even the tactile quality of the paper is satisfying and comforting to our touch. A book is a package; a collection of interrelated pages, each one an integral part of the whole. Like the plates inside a car battery the pages in a book work in unison to generate a charge; a current of thought running from cover to cover.

The printed book is magnificently simple. There is never any worry that the words on the page might suddenly blink-out or slip away into cyberspace. Like us, printed books inhabit the real world.

Chapter 2: *A Printing Revolution*

This illustration from a 19th century schoolbook was accompanied by a fanciful description of the voices in Gutenberg's head:

One night as he sat thinking about his press, he thought he heard two voices.

The first voice said: "Happy, happy man!" "Go on with your great work, and be not discouraged. In the ages to come, men of all lands will gain knowledge and become wise by means of your great invention. Books will multiply until they are within the reach of all classes of people. Every child will learn to read. And to the end of time, the name of Gutenberg will be remembered."

The second voice said: "Beware! beware! and think twice of what you are doing. Evil as well as good will come from this invention upon which you have set your heart. Instead of being a blessing to mankind, it will prove to be a curse. Pause and consider before you place in the hands of sinful and erring men another instrument of evil."

Gutenberg's mind was filled with distress. He thought of the fearful power which the art of printing would give to wicked men to corrupt and debase their fellow-men. He leaped to his feet, he seized his hammer, and had almost destroyed his types and press when the first voice spoke again, and in accents loud enough to cause him to pause.

"Think a moment," it said. "God's gifts are all good, and yet which one of them is not abused and sometimes made to serve the purposes of wicked men. What will the art of printing do? It will carry the knowledge of good into all lands; it will promote virtue; it will be a new means of giving utterance to the thoughts of the wise and the good."

So (fortunately for us) Gutenberg threw down his hammer and got back to work.

William Prynne
1600 – 1669

An illustration showing author William Prynne locked in a pillory in London in 1633. Notice the hot iron about to be used to brand his face. Prynne is one of many authors who have been persecuted or executed for writing a book contrary to the official view of the government or religious authority.

CHAPTER 3

The Power of Books

"Where they have burned books,
they will end in burning human beings."

~ HEINRICH HEINE ~

he significance of Gutenberg's printing press cannot be overstated. It essentially launched humanity into a new age of information, education and enlightenment for the masses.

Printed books weren't solely responsible for this, of course. A combination of technological and societal changes occurred simultaneously, each reinforcing the other. But there is no question that books played a central role in helping fuel the Renaissance, the Protestant Reformation and the Scientific Revolution.

After Gutenberg introduced the world to movable type, print shops sprang up in cities all over the map. By 1500, only about 50 years after Gutenberg's invention, Europe's presses had already cranked-out several million books.

Books democratized information. They made new ideas, opinions and perspectives available to the general public. Governments and powerful institutions like the Catholic Church were not always happy about this. Information is power; the powerful want to control it. The pages of history are filled with examples of books that have been banned or burned – sometimes along with their authors.

In 1633 a London lawyer named William Prynne wrote a book entitled: *Histriomastix: The Player's Scourge, or*

Actor's Tragedy. To say the book was controversial would be a major understatement.

Although Prynne, a Puritan, apparently wrote the book to criticize the immorality of stage plays and actors, some interpreted it as a disguised criticism of England's government, the Catholic Church and the Queen. The book was promptly burned in a public ceremony.

That was only the beginning of poor Mr. Prynne's problems. He was tried and convicted of sedition. He spent a year imprisoned in the Tower of London before being sentenced to the pillory, an instrument of punishment similar to the stocks. Then both of his ears were cut off. But that wasn't the end of it. He was branded on both cheeks using a red-hot iron with the letters "SL" signifying that he was a Seditious Libeller. And, to make sure he got the point, he was fined a huge sum of money and sentenced to life in prison, all because of a book.

Prynne was actually lucky compared to William Tyndale. In 1536, Tyndale, a priest, was convicted of heresy for translating the New Testament into English so it could be read by commoners. This was apparently so threatening to the Church that authorities felt it necessary to kill Tyndale twice. He was strangled *and* burned at the stake.

Nearly 500 years later, books are still igniting outrage. Take for example the 1988 novel, *The Satanic Verses* by Salman Rushdie. The book caused international controversy. Violent demonstrations against the book resulted in several deaths. It was banned in many countries. In Venezuela, owning or reading it carried a penalty of 15 months imprisonment. Iran's leader called for Rushdie's execution. An Iranian charity offered a million dollars for his murder. Rushdie went into hiding.

Meanwhile, the same book became a best-seller in

Famous Banned Books

These are just a few of the famous books that have been banned because of their religious, social, political or sexual content.

The Adventures of Huckleberry Finn, Mark Twain, 1884

The Autobiography of Malcolm X, Malcolm X and Alex Haley, 1965

Beloved, Toni Morrison, 1987

Bury My Heart at Wounded Knee, Dee Brown, 1970

The Call of the Wild, Jack London, 1903

Catch-22, Joseph Heller, 1961

The Catcher in the Rye, J.D. Salinger, 1951

Fahrenheit 451, Ray Bradbury, 1953

Fifty Shades of Grey, E.L. James, 2012

For Whom the Bell Tolls, Ernest Hemingway, 1940

Gone With the Wind, Margaret Mitchell, 1936

The Grapes of Wrath, John Steinbeck, 1939

In Cold Blood, Truman Capote, 1966

Leaves of Grass, Walt Whitman, 1855

Moby-Dick, Herman Melville, 1851

The Red Badge of Courage, Stephen Crane, 1895

The Scarlet Letter, Nathaniel Hawthorne, 1850

Sexual Behavior in the Human Male, Alfred C. Kinsey, 1948

A Streetcar Named Desire, Tennessee Williams, 1947

To Kill a Mockingbird, Harper Lee, 1960

Uncle Tom's Cabin, Harriet Beecher Stowe, 1852

Where the Wild Things Are, Maurice Sendak, 1963

Europe and the United States. In England it won a literary prize for best novel. These diametrically opposed reactions to a book of fiction are reminders that a book is not necessarily inherently good or bad. It depends on who is evaluating its content.

These stories and many others like them illustrate the tremendous power a book can pack. For centuries books have been banned and burned; their authors imprisoned, tortured and executed. At the same time, books have been revered, quoted, collected, gilded, and sworn upon as gospel. A book may be a mere pound of paper but the power generated between its covers can be immense.

Books Transport Us

Books are unique objects. They have the ability to transport thoughts across time (and space). As we read, we are transported as well.

We can open a book and be inspired by the thoughts of an ancient philosopher or enjoy the musings of a poet from the Middle Ages. We might gain perspective from historical accounts of triumphs and tragedies, or find hope in the words of a prophet. Nothing can do this quite like a book.

Of course, books are not simply repositories of knowledge. We read them for entertainment too. Fiction transports us across time in ways nonfiction can't. Historical fiction breathes life into history. Science fiction can carry us into the future or to an alternative version of the past or present. And intentionally or not, each book reflects the time period in which it was written. Few things in our lives can connect us with the human narrative, past, present and future, like the printed book.

Metaphorically Speaking...

Throughout history, writers (including this writer) have tried to characterize the essence and impact of books by using metaphors and similes. Books have been called compasses, portals, time capsules, time machines, windows to the past, doorways to knowledge, etc. They have been likened to lighthouses in a sea of darkness, bees that carry pollen from one mind to another, gardens, tickets and telescopes. Each of these is valid but none by itself adequately expresses the multidimensional quality of books and the immeasurable effects they have on culture. Fiction author Stephen King once described books as: "a uniquely portable magic." That's probably as good a description as any.

Paradigm Shift

There are some obvious comparisons that can be drawn between the invention of the printing press in the mid-15th century and the advent of the Internet only a couple of decades ago. Each of these technologies created a paradigm shift in the way information and ideas are shared.

Both the printing press and the Internet have made information more available to the general public. As we have seen, not everyone considers this a good thing. The persecution of Prynne, Tyndale and Rushdie are just a few examples of this. Although authors aren't getting burned at the stake these days (we hope), the resistance to the free flow of information is still strong in many places. Like books, the Internet is very threatening to oppressive regimes. Today, several countries censor or block access to the Internet.

We've had 500 years to observe the effects printed books have had on the world. There are good books and bad but, in general, books have had an overwhelmingly positive effect. The Internet is new. Its effects and consequences are still unfolding. It too can be good and bad, but if history repeats itself, this emerging technology will ultimately prove to be a positive force in our world.

There is another important connection between books and the Internet: the rise of print-on-demand technology *(see page 29)*. Because of the symbiotic relationship between POD books and the Internet, it wouldn't be an exaggeration to say that the Internet has become part of the Printing Revolution.

The Internet and POD book technology are evolving simultaneously. POD makes it economically practical for any author to publish his or her own book. This in itself is significant. But for the first time in history, self-published authors are able to reach a worldwide audience via the Internet. The power of POD, combined with the power of the Internet, is pumping new life into books. This applies to e-books as well. Gutenberg's technology (albeit highly evolved), and the Printing Revolution, live on.

Gutenberg's Legacy

Imagine an immense bookcase stretching as far as the eye can see. On the shelves of this imaginary bookcase are copies of every book ever published; the accumulated literature of mankind. Each of these books is like a page from our collective narrative. It takes all of them together to tell our whole story. It's a story that is still being written. Everyday, new books are added to the shelves.

Today, we have access to incredible electronic resources. Information on any subject is available at our fingertips. Even so, you are now reading a printed book. Books are still as relevant as they were 500 years ago. This is the legacy of Johannes Gutenberg – Man of the Millennium.

Postscript

Gutenberg *vs.* Coster

A bronze statue in the public square of the city of Haarlem in the Netherlands, depicts the first man to print books using movable type. But it's not a statue of Gutenberg. It's the Netherlands' own Laurens Janszoon Coster (or Koster).

Laurens Coster

An account written several decades after Coster's death claims he was working on the same kind of printing process as Gutenberg at roughly the same time. He was apparently trying to produce metal type using a sand-casting method. Coster was born about twenty-five or thirty years before Gutenberg. He died in a plague around 1440, a decade before Gutenberg printed his bibles.

The statue of Coster holds the letter "A" in its outstretched hand, signifying Coster's invention of movable type. However, there are no known books that can be positively attributed to him.

It's entirely possible that Coster did achieve some success in casting and printing with movable type but, unless definitive evidence of his work comes to light, Gutenberg will likely remain the "Father of Printing."

About the Author

Aaron Keirns is a writer, book designer and teacher with a lifelong interest in history. He worked in the printing and  publishing industry for more than 35 years and has experience in a variety of areas, including layout & design, typography, illustration, photography and journalism. He holds a degree in Anthropology from The Ohio State University and has taught college courses in digital media design. Aaron continues to write and design books on a variety of topics. For more information about his books, visit: www.littleriverpublishing.com.

Acknowledgments

While doing research for this book, the author visited Gutenberg's hometown of Mainz, Germany, and toured the Gutenberg Museum located there. The museum contains a wealth of information about Gutenberg's life and work as well as excellent exhibits on the evolution of printing. Thank you to the museum for generously contributing several of the images used in the book.

Thank you to Nathan J. Keirns, M.A., and Debora M. Tussey for their excellent editorial input, proofreading and encouragement during this project.

Thanks also to:

The Columbus Ohio Metropolitan Library
The Public Library, City of Mainz, Germany
The Public Library of Mount Vernon and Knox Co., Ohio
The Public Library of Newark and Licking Co., Ohio

Glossary

Font
Traditionally, an assortment of type in one style, size and weight (e.g. 12 point Garamond Bold). Today, the word "font" is often used interchangeably with the word "typeface."

Illuminated Manuscript
A manuscript or book supplemented with hand-drawn capital letters, flourishes and other decorative elements.

Justified Type or Text
A column of type aligning flush on both left and right margins.

Letterpress
The process of printing from inked type in relief.

Linotype
A brand of typesetting machine, actuated by a keyboard, that casts lines or "slugs" of type in lead.

Matrix
A mold (mould) used for casting movable type.

Movable Type
Individual letters, typically cast in metal, that can be arranged into words, sentences, etc. for use in letterpress printing.

Phototype
Type reproduced photographically and used in conjunction with light-sensitive printing plates.

Typeface
A complete assortment of type in one style (e.g. Garamond). Today "typeface" and "font" are often used interchangeably.

Typesetting
The process of assembling movable type into words, sentences, etc., in preparation for printing.

Vellum
A type of parchment typically made from the skin of calves.

Woodcut
A carved block of wood from which prints are made. Woodcuts are often illustrations but can contain text or decorative elements.

Colophon

A colophon is a brief description, often placed at the end of a book, listing facts about the printing and publishing of the book. It was a common element of some early books and often included the publisher's insignia or logo. Sometimes the type of the colophon was set in a decorative manner and incorprated an illustration. The practice has been revived from time to time and seems particularly appropriate for this book.

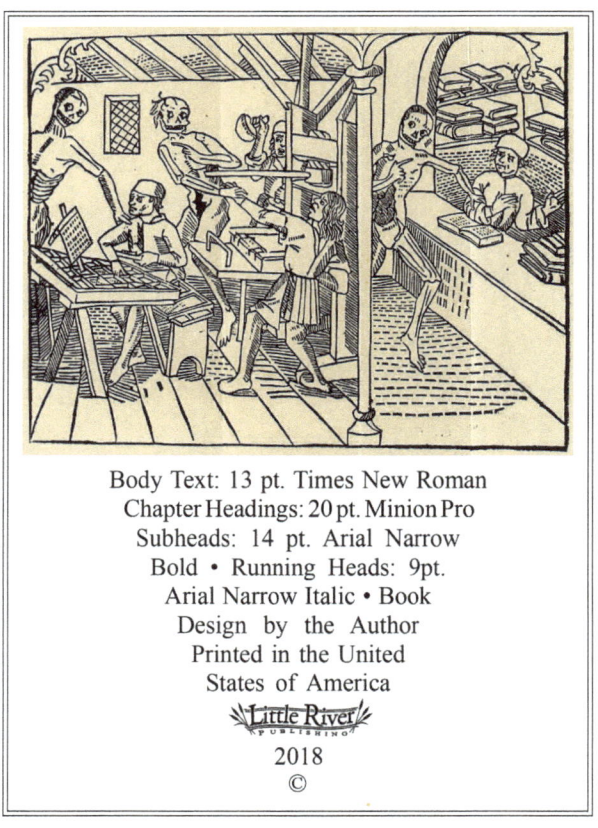

Body Text: 13 pt. Times New Roman
Chapter Headings: 20 pt. Minion Pro
Subheads: 14 pt. Arial Narrow Bold • Running Heads: 9pt. Arial Narrow Italic • Book Design by the Author
Printed in the United States of America
Little River
2018
©

Note: The woodcut illustration above, entitled "Danse Macabre" (Dance of Death) by Mathias Huss, 1499, shows Death coming for the printers. This theme occurs several times in medieval art and is an expression of the fear and anxiety caused by the introduction of the printing press. Essentially, it's saying that all printers are going to Hell. This is also believed to be the earliest illustration of a printing press.

Index

B
Banned books, 37
Bible, 3, 7-8, 13-14, 21, 24-27
Bindery, 25
Block book, 13

C
Colophon, 44
Composing stick, 12, 18
Compositor, 18
Coster (Koster), 41

E
E-books (electronic books), 29, 31-32, 40

F
Font, 43
Fry, Stephen, 13, 32
Fust, Johann, 8

I
Illumination, 25
Ink, 3, 6, 20-21

J
Justified type, 24, 43

K
King, Stephen, 39

L
Letter of Indulgence, 8
Letterpress, 19, 43
Linotype, 28, 43

M
Mainz, 1, 3, 8, 10-11, 42
Matrix, 14-17, 28, 43
Movable type, 3, 4, 14, 19, 21, 28, 41, 43

P
Paper, 20-21, 24
Phototype (phototypesetting), 29, 43
POD (print-on-demand), 29, 40
Prynne, William, 34-36, 39

R
Rusdie, Salman, 36, 39

S
Scriptorium, 2

T
Tyndale, William, 36, 39
Type Caster, 15-17
Typeface, 43
Typesetting, 18, 29, 43

V
Vellum, 20, 24, 43

www.ingramcontent.com/pod-product-compliance
Lightning Source LLC
Chambersburg PA
CBHW041812040426
42450CB00001B/9